Library of Congress Cataloging-in-Publication Data

Coldrey, Jennifer.
Strawberry/Jennifer Coldrey: photographs by George Bernard.
p. cm.–(Stopwatch books)
Includes index.
Summary: Photographs, drawings, and text on two levels of difficulty describe how the strawberry plant produces juicy strawberries and sends out runners to start new plants.
1. Strawberries–Juvenile literature. 2. Strawberries–Development–Juvenile literature. [1. Strawberries.]
I. Bernard, George, 1949- ill. II. Title. III. Series.
SB385.C58 1988
634'.75–dc19 88-18344
ISBN 0-382-09802-1. CIP
ISBN 0-382-09801-3 (lib. bdg). AC

First published by A & C Black (Publishers) Limited
35 Bedford Row, London WC1R 4JH

Adapted and published in the United States in 1989
by Silver Burdett Press, Englewood Cliffs, New Jersey
U.S. project editor: Nancy Furstinger

Acknowledgments
The artwork is by Helen Senior

Strawberry

Jennifer Coldrey
Photographs by George Bernard

Silver Burdett Press • Englewood Cliffs, New Jersey

Here is a bowl of strawberries.

How do you like to eat strawberries?
In a cake? Or with cream? Or just on their own?
Strawberry plants grow in gardens and fields, like this.

Can you see the ripe strawberries?

This book will tell you how strawberries grow.

The strawberry plant has roots.

The strawberry plant rests through the winter.
When spring comes, new leaves start to grow.
The plant puts down more roots into the soil.

Here are some of the roots shown up close.

Can you see the tiny hairs on these roots?
The hairs take in water from the soil and carry it to the rest of the plant. The plant needs water and nutrition from the soil to live and grow.

Flower buds grow on the plant.

As the days grow longer and the weather gets warmer, the strawberry plant gets bigger. The leaves spread out to catch as much light as possible.

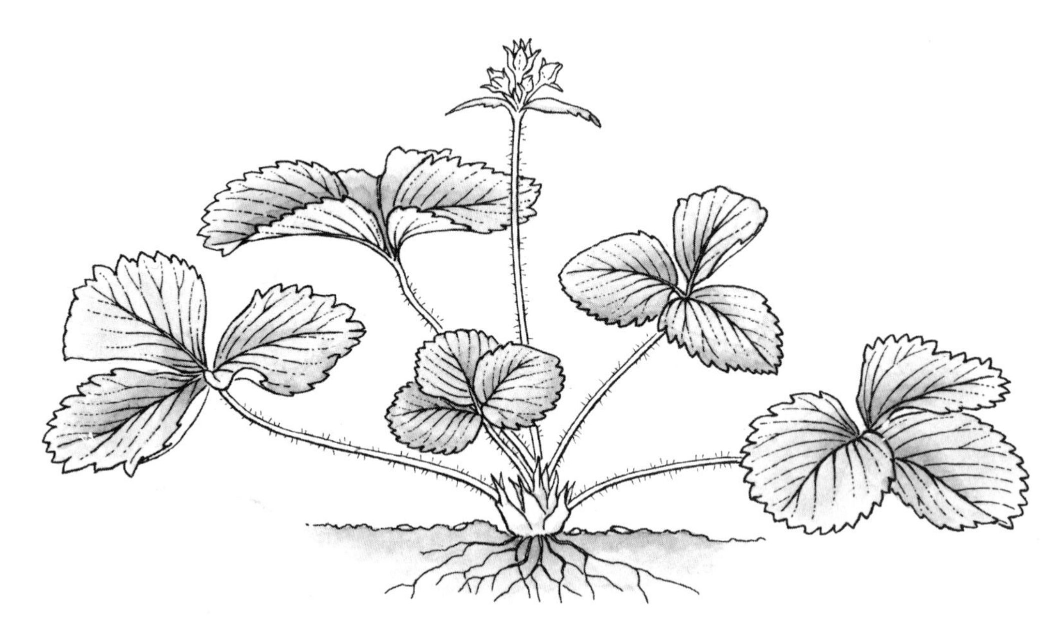

The leaves use sunlight to help make food for the plant.

In early spring the strawberry plant grows flower buds. Look at the big photograph. The buds are covered by green sepals (tiny leaves). Soon the flowers will open, one by one.

Insects visit the flowers.

Each flower has five petals around the outside. In the middle there is a cushion of tiny yellow stalks.

Can you see the longer stalks with brownish-yellow tops? They are covered with yellow dust called pollen.

Look at the big photograph. This bumblebee is visiting the flower to look for food. As it crawls over the flower, some pollen may brush onto its body.

A tiny strawberry starts to grow.

When the insect visits another strawberry flower, some pollen may brush off its body onto the middle of the flower. Then a tiny strawberry may start to grow.

The petals fall off the flower, and the middle begins to swell up. This flower has been cut in half.

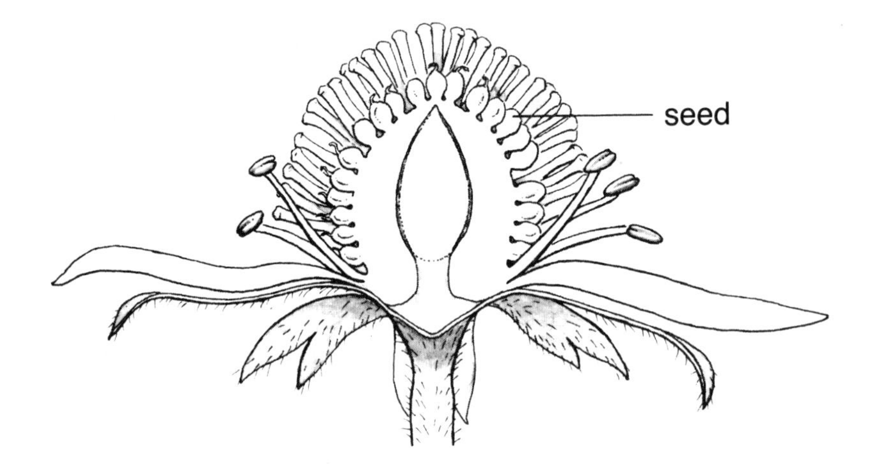

Can you see the tiny seeds around the middle of the flower? They are the seeds of the strawberry.

Look at the big photograph. How many strawberries are growing on this plant?

The strawberry grows bigger and turns red.

The seeds are on the outside of the strawberry. As it grows and swells, the seeds grow farther apart. At first the strawberry is green. Look at the small photograph.

This strawberry has been growing for three weeks. It is still quite hard and it is not ready to eat.

As the strawberries grow, they slowly turn cream color, then pink, then red. Look at the big photograph. These strawberries started growing at different times.

This strawberry is ready to eat.

When the strawberry turns red, it is ripe and ready to eat. This photograph shows the skin of the strawberry up close.

Each seed is sunk into a little dip in the skin.

Look at the big photograph. Each seed is joined to the middle of the strawberry by a pale line. These lines are the veins that carry food to the seeds.

The strawberries are eaten.

Most strawberries are eaten by people, or by birds and other animals. Look at the big photograph. A field mouse is eating this ripe strawberry.

If the strawberry is not eaten, it will grow old and rot. These strawberries have mold growing on them.

When the strawberries rot, the seeds may fall onto soil. Some may start to grow into new strawberry plants.

The strawberry plant grows runners.

During the summer, when the fruits are growing, long creeping stems start to grow from the strawberry plant. These stems are called runners.

Some of the runners can grow as long as your arm. At the tip of each runner there is a bud. It is protected by tiny hairs.

Look at the big photograph. Soon the bud opens and tiny leaves begin to unfold. A new strawberry plant starts to grow.

The young plant grows bigger.

After a few days the young strawberry plant has grown more leaves. The runner carries food to the new plant. Look at the big photograph. Tiny roots are starting to grow.

The roots will grow down into the soil. They will hold the plant in the ground, and draw up food and water for the young plant.

The strawberry plant puts out lots of runners. Each one grows into a new strawberry plant.

The plant rests through the winter.

By autumn the young strawberry plant has grown bigger. Its roots grow deep into the soil and it has lots of new leaves. Even if the runner is cut, the plant can now make its own food.

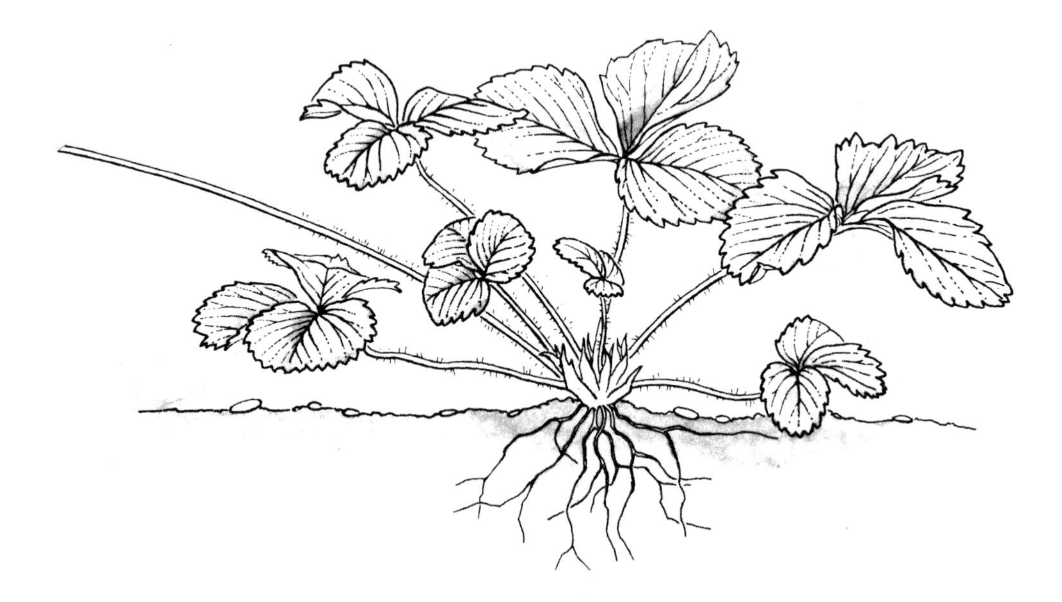

In winter, the strawberry plant stops growing. It rests until the spring. When the weather is warmer, new leaves and flower buds will start to grow.

What do you think will happen then?

Do you remember how a strawberry plant grows?
See if you can tell the story in your own words.
You can use these pictures to help you.

Do you know a garden where strawberries grow?
Count how many runners grow from one plant.

Index

This index will help you to find some of the important words in this book.